ENERGY POLICIES, POLITICS AND PRICES

ALTERNATIVE FUEL INVESTMENTS BY THE DEPARTMENT OF DEFENSE

ASPECTS AND CONSIDERATIONS

ENERGY POLICIES, POLITICS AND PRICES

Additional books in this series can be found on Nova's website under the Series tab.

Additional e-books in this series can be found on Nova's website under the e-book tab.

ENERGY POLICIES, POLITICS AND PRICES

ALTERNATIVE FUEL INVESTMENTS BY THE DEPARTMENT OF DEFENSE

ASPECTS AND CONSIDERATIONS

LOREN MURPHY
EDITOR

New York

NOTICE TO THE READER

Library of Congress Cataloging-in-Publication Data

ISBN: 978-1-63484-549-6

Published by Nova Science Publishers, Inc. † New York

CONTENTS

PREFACE

The Department of Defense (DOD) is the single largest consumer of energy in the federal government, spending billions of dollars annually on petroleum fuels to support military operations. One of DOD's strategic operational energy goals is to expand its energy supply options. Investing in alternative fuels—liquid fuels, derived from non-petroleum feedstocks, whose use does not necessitate any modifications to platforms and equipment—represents one means of potentially achieving this goal. This book reviews the extent to which DOD has purchased alternative fuels, and has demonstrated these fuels can meet its safety, performance, and reliability standards; has a process for purchasing alternative fuels for military operations that takes into consideration any cost differences between alternative and conventional fuels; and has used the DPA authorities to promote the development of a domestic biofuel industry.

In: Alternative Fuel Investments …
Editor: Loren Murphy

ISBN: 978-1-63484-549-6

Chapter 1

DEFENSE ENERGY: OBSERVATIONS ON DOD'S INVESTMENTS IN ALTERNATIVE FUELS*

United States Government Accountability Office

WHY GAO DID THIS STUDY

DOD is the single largest consumer of energy in the federal government, spending billions of dollars annually on petroleum fuels to support military operations. One of DOD's strategic operational energy goals is to expand its energy supply options. Investing in alternative fuels—liquid fuels, derived from non-petroleum feedstocks, whose use does not necessitate any modifications to platforms and equipment—represents one means of potentially achieving this goal.

GAO was asked to examine aspects of DOD's investment in alternative fuels. GAO reviewed the extent to which DOD (1) has purchased alternative fuels, and has demonstrated these fuels can meet its safety, performance, and reliability standards; (2) has a process for purchasing alternative fuels for military operations that takes into consideration any cost differences between alternative and conventional fuels; and (3) has used the DPA authorities to promote the development of a domestic biofuel industry.

* This is an edited, reformatted and augmented version of a United States Government Accountability Office publication, No. GAO-15-674, dated July 2015.

GAO reviewed past alternative and conventional petroleum fuel procurements, as well as statutes, regulations, and DOD guidance related to fuel purchases. Also, GAO reviewed various documents on biofuel projects initiated under the DPA authority and interviewed cognizant DOD officials involved with purchasing and using fuel and administering the biofuel projects.

GAO is not making recommendations in this report. DOD provided technical comments on the findings, which GAO has incorporated where appropriate.

What GAO Found

The Department of Defense (DOD) has purchased small quantities of alternative fuels—jet and naval distillate (known as F-76, to power ships)—for testing and demonstration purposes, but has not done so yet for military operations. DOD's testing process validates the ability of alternative fuels to meet safety, performance, and reliability standards for military equipment and platforms. From fiscal years 2007 through 2014, DOD purchased about 2.0 million gallons of alternative fuel for testing purposes, at a cost of about $58.6 million. Over the same period, it purchased about 32.0 billion gallons of petroleum fuel at a cost of about $107.2 billion. DOD has approved alternative fuels made from two production processes for use in certain items and is continuing to test others.

DOD is currently required by law to ensure alternative fuel purchases for operational purposes are cost-competitive with conventional fuels and has a standard process to purchase large-scale volumes of all fuels. Proposals are evaluated according to technical acceptability and price. To help the Navy purchase alternative jet and naval distillate fuels blended with conventional fuels, the Department of Agriculture plans to provide funding directly to alternative fuel vendors that meet certain requirements and receive awards from DOD. These funds are intended to defray some of the alternative fuel producer's extra costs— such as costs of domestic feedstocks. Per DOD, no alternative fuel vendors have received awards so none of these funds have been paid out yet.

DOD has used financial incentives provided for by Title III of the Defense Production Act (DPA) to help facilitate the development of commercially viable plants for producing biofuels for the military and commercial sectors. To date, DOD has used this authority for two ongoing projects: Bio-Synthetic

Paraffinic Kerosene and Advanced Drop-In Biofuels Production Project and the federal government's cost share for these projects was about $234.1 million.

ABBREVIATIONS

DLA-E	Defense Logistics Agency-Energy
DOD	Department of Defense
DPA	Defense Production Act

* * *

July 27, 2015

The Honorable Mac Thornberry
Chairman
Committee on Armed Services
House of Representatives

Dear Mr. Chairman:

The Department of Defense (DOD) is the single largest consumer of energy in the federal government. A majority of DOD's energy usage goes toward its operational energy needs—that is, the energy required for training, moving, and sustaining military forces and weapon platforms for military operations.[1] DOD largely depends on petroleum fuels for military operations. In fiscal year 2014, the Departments of the Navy, Air Force, and Army purchased about 3.8 billion gallons of petroleum fuel and other fuel products at a total cost of about $14.4 billion (not adjusted for inflation).[2] Given the realities of global oil markets, according to DOD's Operational Energy Strategy, a disruption of oil supplies is plausible and increasingly likely in the coming decades and the volatility of oil prices will continue to pose a budgetary challenge.[3]

According to DOD, one of the department's strategic operational energy goals is the expansion of its energy supply options. One means of potentially achieving this goal is DOD's investment in alternative fuels— an investment that DOD cites as a prudent insurance policy against future oil supply disruptions and high prices.[4] DOD's primary alternative fuels goal, according to DOD policy, is to ensure operational military readiness, improve battlespace effectiveness, and further the flexibility of military operations through the ability to use multiple, reliable fuel sources.[5] Further, DOD policy indicates alternative fuels can serve as a mechanism for mitigating anti-access/area denial[6] effects, and for enabling flexibility in supply chain logistics. In addition, the 2014 Quadrennial Defense Review notes that energy improvements enhance range, endurance, and agility, particularly in the future security environment where logistics may be constrained.[7]

You asked us to examine aspects of DOD's investment in alternative fuels. This report discusses the extent to which DOD (1) has purchased alternative fuels and has demonstrated that these fuels can meet its safety, performance, and reliability standards; (2) has a process for purchasing alternative fuels for military operations that takes into consideration any cost differences between alternative and conventional fuels; and (3) has used the Defense Production Act (DPA) authorities to promote the development of a domestic biofuel industry.[8]

To determine the extent to which DOD has purchased alternative fuels, and has demonstrated that these fuels can meet its safety, performance, and reliability standards, we obtained and analyzed the military departments' alternative fuel purchase data—including quantities, costs, and funding sources—covering fiscal years 2007 through 2014. Through reviews of DOD guidance, reports, publications, and briefings and interviews with DOD officials, we obtained information about the reason(s) why the military departments purchased alternative fuels and how these fuels were put to use. To assess the reliability of the data, we reviewed the information and met with the DOD officials involved with purchasing alternative fuels about the source of the data. We concluded that the purchase data are of sufficient reliability to describe the order of magnitude of DOD's funding toward alternative fuels. In addition, we obtained data on the military departments' conventional petroleum and other fuel product purchases from fiscal years 2007 through 2014 based on database queries that DOD officials conducted from March through April 2015. To assess the reliability of the data, we analyzed the data and gathered information on the source of the data and the functionality of the database—including data entry and monitoring processes. We also reviewed any inconsistent information (e.g., negative values) with DOD officials to understand any issues identified with the data. Based on an interview with DOD officials and our review, we determined that the data are of sufficient reliability to provide context about the quantities and cost of petroleum purchases within DOD. We selected the fiscal year 2007 through 2014 time frame for each of the data series above after assessing their availability and reliability to maximize the amount of data available for us to provide information and context.

To determine the extent to which DOD has a process for purchasing alternative fuels for military operations that takes into consideration any cost differences between alternative and conventional fuels, we reviewed relevant statutes, regulations, and DOD directives and other guidance that provide direction on buying fuel. In addition, we obtained and analyzed documentation about an interagency initiative related to alternative fuel purchases. We reviewed briefings and fuel contract solicitations for information about how DOD would address any cost differences between alternative and conventional fuels during the fuel purchase process. We also interviewed DOD officials responsible for and involved with fuel acquisitions.

To determine the extent to which DOD has used the DPA authorities to promote the development of a domestic biofuel industry, we reviewed relevant statutes, a DOD directive, and other guidance generally related to the DPA authorities. We interviewed DOD officials who implement and oversee the implementation of the authorities about the instances where they have been used to promote the development of a domestic biofuel industry. We obtained and reviewed documentation on related biofuel projects that DOD has initiated under the authorities including a memorandum of understanding, budget estimate submissions for fiscal years 2011 to 2015, project documentation submitted to congressional committees, project solicitation and agreement documents, and related statutes—such as related provisions in authorization and appropriations acts. In addition, when we met with DOD officials, we obtained information on how ongoing biofuel projects were initially evaluated, funded, and monitored.

We conducted this performance audit from September 2014 to July 2015 in accordance with generally accepted government auditing standards. Those standards require that we plan and perform the audit to obtain sufficient, appropriate evidence to provide a reasonable basis for our findings and conclusions based on our audit objectives. We believe that the evidence obtained provides a reasonable basis for our findings and conclusions based on our audit objectives.

BACKGROUND

Types of Fuel Used in Military Operations

Military operations generally rely on petroleum-based fuels that power communication equipment, forward deployed bases, tactical and combat ground vehicles, aircraft, naval vessels, and other platforms. For military operations, DOD primarily uses jet and naval distillate fuels. With regard to jet fuel, until recently DOD predominately used two types: Jet Propellant-8 and Jet Propellant-5. Jet Propellant-8 is the primary fuel used in DOD land-based aircraft, tactical and combat ground vehicles, and ground support equipment.[9] Aircraft, ground vehicles, and support equipment deployed aboard maritime vessels (such as aircraft carriers) use Jet Propellant-5 since this jet fuel—which is stored in large quantities—is less combustible than other types of jet fuel, which is important for fire safety reasons. The standards for these two fuels are listed in the DOD technical fuel specification documents maintained by the Departments of the Air Force and Navy, respectively, and approved for use by all DOD departments and agencies.[10] Jet Propellant-8 fuel is similar to commercial jet fuels—known as Jet A and Jet-A-1.[11] The standards for these fuels are listed in the technical fuel specification document issued by ASTM International.[12] As of the end of calendar year 2014, DOD reported completing conversion from Jet Propellant-8 fuel to Jet A—with the inclusion of specific additives for military unique requirements—at military installations within the United States. DOD uses naval distillate fuel, known as F-76, to power nonnuclear ships. This fuel can be burned in shipboard boilers, diesel engines and gas turbines. The standards for this fuel are listed in the DOD technical fuel specification document maintained by the Department of the Navy.[13] Unlike the case of jet fuel, according to DOD officials, there is no commercial equivalent that meets the Navy's maritime needs.[14]

DOD Guidance on Alternative Fuels

DOD Directive 4180.01, DOD Energy Policy, among other things, establishes that DOD will diversify and expand its energy supplies and sources, including alternative fuels.[15] The Energy Policy assigns responsibilities for various matters, including the following:

- The Assistant Secretary of Defense for Operational Energy Plans and Programs 16 is to develop policy and guidelines and provide oversight for development, certification, qualification, field demonstration, and ongoing purchases of alternative fuels for operational platforms in accordance with the U.S. Code;
- The Director, Defense Logistics Agency is to (1) manage energy commodities and related services 17; and (2) provide energy expertise to support the qualification of alternative fuels and support field demonstration activities; and
- The Secretaries of the military departments are to develop and implement doctrine, guidance, and strategies consistent with the directive and implementing instructions.

DOD has also recently issued DOD Instruction 4140.25, DOD Management Policy for Energy Commodities and Related Services, which among other things, establishes DOD policy that:

- alternative fuels are obtained using the DOD's standard fuel procurement programs;
- alternative fuels for operational purposes are purchased when cost-competitive with traditional fuels and when qualified as compatible with existing equipment and infrastructure; and
- fuel systems are qualified to use available commercial-type fuels, including alternative fuels.[18]

DOD's Alternative Fuels Policy for Operational Platforms lists the department's primary alternative fuels goal as, among other things, furthering flexibility of military operations through the ability to use multiple, reliable fuel sources.[19] Additionally, it articulates considerations for the department's investment in alternative fuels, to include: increasing DOD's resilience against strategic supply disruptions, reducing the effect of petroleum price volatility, and increasing fuel options for operational commanders. This policy indicates alternative fuels can serve as a mechanism for mitigating anti-access/area denial effects, and for enabling flexibility in supply chain logistics. The policy stresses that the desired end-state of investments in alternative fuels is operational military readiness and battlespace effectiveness.

DOD's Operational Energy Strategy and related Operational Energy Implementation Plan 20 identify the high-level goal of expanding DOD's operational energy supply options. Promoting the development of alternative fuels—in the form of testing and approving them for use by existing military platforms, and helping to catalyze a competitive biofuels industry—constitutes one means for achieving this goal.

Each of the three military departments has energy guidance documents that address alternative energy or alternative fuels.[21] Two of DOD's military departments—the Navy and the Air Force—have also established usage goals for alternative fuels. The Department of the Navy's guidance sets a goal of deriving 50 percent of total energy consumption from alternative sources—including alternative fuels—by 2020, which, according to Navy estimates, would require using about 336 million gallons of alternative fuels (both naval distillate and jet fuels) annually by 2020. In addition to setting quantitative goals, the guidance established a goal of demonstrating (which the Department of the Navy completed in July 2012) and deploying (by 2016) the Great Green Fleet—that is, ships and aircraft fueled by alternative fuels and other alternative energy sources or utilizing other energy conservation measures.[22] The Department of the Air Force's guidance includes a goal of increasing, to 50 percent of total consumption, the use of cost-competitive drop-in alternative jet fuel blends for non-contingency operations by 2025. Although the Department of the Army uses jet fuel in its tactical and combat ground vehicles, aircraft, and other ground support equipment (such as generators) and engages in efforts to test and approve alternative jet fuel for use in these platforms, the Army does not have specific alternative fuel usage goals in its energy guidance. For additional details about each military department's guidance as related to alternative fuels, see appendix I.

GAO's Prior Work

In May 2014, we reported that broad national strategies promote the development of a variety of alternative fuels—including alternative jet fuel—to help achieve national goals, such as securing energy independence, fostering economic development, and reducing greenhouse gas emissions.[23] Further, we highlighted the efforts of four select federal agencies—including DOD—to sponsor research that specifically targets alternative jet fuel development or provide direct support for its future commercial production, or both. For example, we described multiple non-DOD research and development projects that provide federal support for helping to develop technologies and processes necessary for the commercial production of biofuels. Regarding DOD, we noted DOD's activities to test and approve alternative jet fuel, in particular. In addition, we underscored that while federal government activities help to address the main challenge of alternative jet fuel's price-competitiveness, it is market factors that affect the long-term commercial viability of alternative jet fuels.

Defense Production Act

The Defense Production Act (DPA) generally provides the authority to, among other things, expedite and expand the supply of critical resources from the U.S. industrial base to support the national defense.[24] Title III of the Act—Expansion of Productive Capacity and Supply—allows military and civilian agencies to provide a variety of financial incentives to domestic firms to invest in production capabilities, so as to ensure that the domestic industrial and technological base is capable of meeting the national defense needs[25] of the United States.[26] Among other provisions, Title III authorizes the president to provide for the following in order to create, maintain, protect, expand, or restore domestic industrial base capabilities essential for the national defense:

- purchases of or commitments to purchase an industrial resource or critical technology item;
- encouragement of exploration, development, and mining of critical and strategic materials, and other materials;
- development of production capabilities; and
- increased use of emerging technologies in security program applications and rapid transition of emerging technologies.

Use of the authorities is subject to conditions and requirements established by statute. For example, prior to using the above DPA authorities, the president must determine that the industrial resource, material, or critical technology item is essential to the national defense, that U.S. industry cannot reasonably be expected to provide the capability in a timely manner, and that purchases, purchase commitments, or other actions are the most cost effective, expedient, and practical alternative method for meeting the need.[27] According to DOD officials, the focus of Title III is to establish commercially viable industrial capabilities that will continue to prosper after federal government assistance ends.

The DPA fund manager is the Secretary of Defense. Within DOD, the Under Secretary of Defense for Acquisition, Technology, and Logistics provides guidance to implement the DPA and monitors the Title III program.[28] The Air Force serves as the Executive Agent for DOD's Title III program and maintains a program office to manage and administer aspects of individual Title III projects. Program Office activities include conducting market research and analysis when assessing potential Title III projects; monitoring the technical and business performance of firms receiving Title III financial incentives; and overseeing aspects of contracting for Title III projects.

DOD PURCHASES ALTERNATIVE FUELS TO TEST AND VALIDATE THAT THE FUEL CAN MEET SAFETY, PERFORMANCE, AND RELIABILITY STANDARDS

DOD Has Purchased Small Quantities of Alternative Fuels

DOD has purchased small quantities of alternative fuels for research, development, and demonstration purposes but not large quantities for military operations yet. DOD's energy and alternative fuels guidance discusses the research and development aspects of alternative fuels—to include testing and approving fuels, as well as demonstrating their use in an operational environment—and DOD's Operational Energy Strategy lists conditions for investment in the research, development, testing, and evaluation of alternative fuels.[29] The guidance notes that DOD is currently purchasing alternative fuels for testing purposes, at a premium price— that is, prices higher than those for conventional fuels. The marginal unit cost of producing a commodity at small scale with new processes being researched and developed is typically much higher than the cost of producing the same or similar commodities using existing large-scale commercial production facilities.

The military departments purchased about 2.0 million gallons of alternative jet and naval distillate fuels from fiscal years 2007 through 2014 to conduct the department's testing, approving, and demonstration activities, at a total cost of about $58.6 million (adjusted for inflation to fiscal year 2015 dollars using the gross domestic product price index).[30] This total amount includes about 450,000 gallons for the Department of the Navy's July 2012 Great Green Fleet demonstration with a group of ships and aircraft fueled by alternative fuels in an operational environment that was part of a larger, biennial multinational maritime exercise, known as the Rim of the Pacific exercise.[31] According to DOD officials, the funding sources for the quantities of alternative fuels purchased include each military department's Research, Development, Test, and Evaluation funds and the Department of the Navy's Operations and Maintenance funds—for the Great Green Fleet demonstration—as well as other funds that DOD identified as being associated with the American Recovery and Reinvestment Act of 2009.

By contrast, over the same period of time, the military departments purchased approximately 32.0 billion gallons of jet and naval distillate conventional petroleum fuel at a total cost of about $107.2 billion (adjusted for inflation to fiscal year 2015 dollars using the gross domestic product price index).[32] Figure 1 shows the total quantities and costs of the military departments' jet and naval distillate alternative and conventional petroleum fuels purchases from fiscal years 2007 through 2014. For more details on the quantity and cost of the military departments' jet and naval distillate alternative and conventional petroleum fuels purchases by each fiscal year, see appendix II.

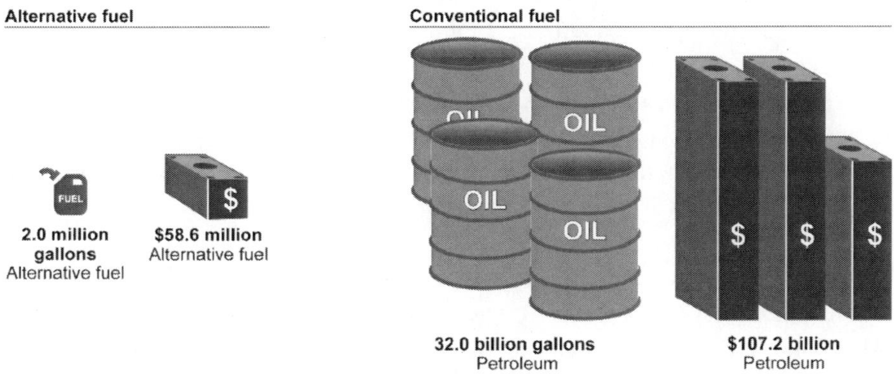

Source: GAO analysis of Department of Defense (DOD) information. I GAO-15-674.

Note: Figure not to scale. The alternative fuel quantity and cost amounts include: Jet Propellant-8, Jet Propellant-5, and Naval Distillate (known as F-76) fuel products. The alternative fuel cost reflects only the cost of fuel. The conventional fuel quantity and cost amounts include: Jet Propellant-8, Jet Propellant-5, Jet A, Jet A-1, and Naval Distillate (known as F-76) fuel products. The petroleum cost reflects both fuel and non-product costs—such as storage and transportation costs—that the military departments paid. All cost data have been adjusted for inflation to fiscal year 2015 dollars using the gross domestic product price index.

Figure 1. Military Departments' Total Alternative and Conventional Fuel Purchases from Fiscal Years 2007 through 2014.

DOD's Testing Process Validates Whether Alternative Fuels Can Meet Safety, Performance, and Reliability Standards of Military Equipment and Platforms

Before any alternative fuel can be used in military operations, DOD tests the fuel to validate whether it can meet unique safety, performance, and reliability standards of military equipment and platforms. These standards reflect the disparate environments in which the military operates—from extreme cold weather to desert geography—and the different types of functionality present in military equipment and platforms—such as flying at high altitudes for military reconnaissance purposes or the afterburner thrust augmentation in military aircraft engines. Two examples of fuel properties important to these standards include a liquid fuel's flash point—the temperature at which existing vapors will combust, or ignite— and its freeze point—the temperature at which it freezes, which affects how it behaves at low temperatures. Jet fuel (specifically Jet Propellant-5) used on military ships is required to have a substantially higher flash point than other jet fuels for safety reasons, since this fuel is stored in large quantities on aircraft carriers and other vessels. A liquid fuel's freezing point potentially can have an effect on certain long-range, high-altitude missions during which extreme cold temperatures are encountered. Requirements for alternative fuels are set out in the relevant DOD technical fuel specification documents.[33] For more details about fuel properties, see appendix III.

The Departments of the Navy, Air Force, and Army test alternative fuels to ensure that they can be used in and on tactical and combat ground vehicles and ground support equipment, ships, aircraft, and fuel distribution systems. The military departments follow a testing and approval process that is similar to that used in evaluating whether to include prospective alternative fuels in the commercial jet fuel standard issued by ASTM International. [34] The departments' testing process captures technical data through laboratory, component, engine, and weapon system platform tests that evaluate the effects of alternative fuels on the performance and reliability of military hardware. The chemical properties of an alternative fuel may be tested in a laboratory using small quantities of fuel—as little as 500 milliliters. As that fuel progresses through the testing process, however, fuel quantity requirements increase. For example, testing alternative fuels in jet engines could require 60,000 gallons of fuel. For more details about the overall testing process, see appendix IV.

Some aspects of the testing process are conducted at DOD laboratories and test facilities, while others are contracted out, according to DOD officials. Throughout the testing process, the military departments share their test data, reports, and expertise within DOD and the federal government and with organizations outside of government—such as ASTM International, equipment manufacturers, and fuel producers. For example, the Department of the Army used test data produced by the Departments of the Navy and Air Force. Consequently, according to DOD officials, they do not duplicate tests previously conducted by another military department; however, when necessary, a department may conduct additional tests if there are fuel properties that are specifically important in certain military applications. In addition, DOD officials stated that they are streamlining the number of specific tests they conduct as they gain more expertise with alternative fuels. Other DOD stakeholders who need to know about fuel issues—such as platform program managers—review the testing results and share their feedback. Once these stakeholders concur that test results demonstrate the prospective alternative fuel meets safety, performance, and reliability expectations and share their approval, the applicable DOD technical fuel specification documents are updated.

DOD Has Approved the Use of Alternative Fuels Made from Two Production Processes and Continues to Test Others

Certain alternative fuels made from the Fischer-Tropsch and Hydroprocessed Esters and Fatty Acids production processes have been tested and approved for use in Navy aviation and ship platforms, Air Force aviation assets, and Army tactical and combat ground vehicles and ground support equipment, but not yet for Army aviation assets.[35] Under the previously mentioned DOD technical fuel specification documents, alternative fuels produced through these two processes are approved for up to a 50 percent blend with conventional fuel. According to DOD officials, the alternative fuels made from these production processes that were used in the testing process included fuels derived from natural gas, coal, and renewable biomass (such as camelina, algal oil, and tallow) feedstock sources.

The military departments continue to have some alternative fuel testing efforts underway. Currently, according to Department of the Army officials, the testing process for alternative fuels made from the two production processes discussed above, as well as fuel made from the Alcohol to Jet production process, for use in Army aviation assets is complete.[36] Further, they stated that the test results are undergoing review in order to decide whether to approve the use of these fuels in Army aviation assets. Also, according to Department of the Army officials, they plan to complete the testing of alternative fuels made from the Alcohol to Jet production process for use in tactical and combat ground vehicles and ground support equipment by the end of calendar year 2015. They stated they plan to start considering, just for these platforms, alternative fuels made from the Synthesized Iso-Paraffins and Catalytic Hydrothermolysis production processes before the end of fiscal year 2015 by purchasing fuel and beginning some testing.[37] According to a Navy official, the testing for alternative fuel made from the Alcohol to Jet and Synthesized Iso-Paraffins production processes in aviation platforms is complete while testing of these fuels in ship platforms is ongoing. In addition, the Department of the Navy has begun testing alternative fuels made from the Catalytic Hydrothermolysis and Hydroprocessed Depolymerized Cellulosic production processes.[38] According to an Air Force official, beyond updating previously conducted tests of alternative fuels made from other production processes—such as Alcohol to Jet— there are no ongoing or planned efforts within the Department of the Air Force to complete additional testing and approval. According to Air Force officials, if other fuel production processes appear likely to become commercially viable, the Air Force will revisit resuming its alternative fuel testing and approval efforts.[39]

As discussed above, DOD recently reported converting from purchasing military-specification Jet-Propellant 8 jet fuel in the United States to purchasing commercial-grade jet fuel—Jet A—with specific additives for military-unique requirements as a means for cost savings and broadening the fuel provider supply pool. While commercial-grade jet fuels blended with alternative fuels are not being produced on a commercial scale in the United States, ASTM International has approved, for commercial aviation, the use of three types of alternative fuels produced through the (1) Fischer-Tropsch and Hydroprocessed Esters and Fatty Acids production processes discussed above for up to a 50 percent blend with conventional fuel and (2) Synthesized Iso-Paraffins production process referenced above for up to a 10 percent blend with conventional fuel. In addition, ASTM International continues to evaluate and consider approving alternative fuels made from other production processes, including those cited above for use in commercial aviation. Consequently, DOD could use alternative fuels in the future via Jet A fuels once those fuels are available on a commercial scale and a widespread basis in the fuel marketplace. According to DOD officials, unless DOD continues to test and approve additional alternative fuels that are being approved by ASTM International, DOD runs the risk of having to develop a separate supply chain for jet fuel—in other words, buying a specialty jet fuel product—as it cannot be assured that commercial-grade jet fuel will meet military safety, performance, and reliability standards.

DOD Has a Standard Process to Purchase All Fuels for Military Operations and Is Currently Required to Ensure Alternative Fuel Purchases for Operational Purposes Are Cost-Competitive with Conventional Fuels

DOD Uses a Standard Process to Purchase Large-Scale Volumes of Fuel for Military Operations

DOD has a standard process in place for purchasing large-scale volumes of fuel, including alternative fuels, for military operations. In support of DOD's large-scale fuel program, the Defense Logistics Agency Energy (DLA-E) activity[40] provides worldwide energy support, including for large-scale fuel purchasing, transportation, and storage for the military and other government customers.[41] As depicted in figure 2 below, DLA-E purchases fuel worldwide in large volumes via four major regions: Inland/East/Gulf Coast/Offshore; Rocky Mountain/West Coast/Offshore; Atlantic/European/Mediterranean; and Western Pacific. DLA-E considers two primary factors—technical acceptability and price—when evaluating fuel vendors' submitted proposals. The fuel must first meet DOD's technical fuel specifications and other technical evaluation factors as part of the consideration. DOD officials indicated that DLA-E typically awards multiple 1-year contracts in these purchase programs. Because the price of energy commodities changes frequently, DOD documents indicate that DLA-E and DOD establish fuel purchase contracts that are tied to market price indicators with fixed margins. The fuel is moved through a commercial distribution system (via tankers, railcars, barges, tank trucks, and pipelines) to intermediary storage locations for redistribution, or directly to the end use military customer.

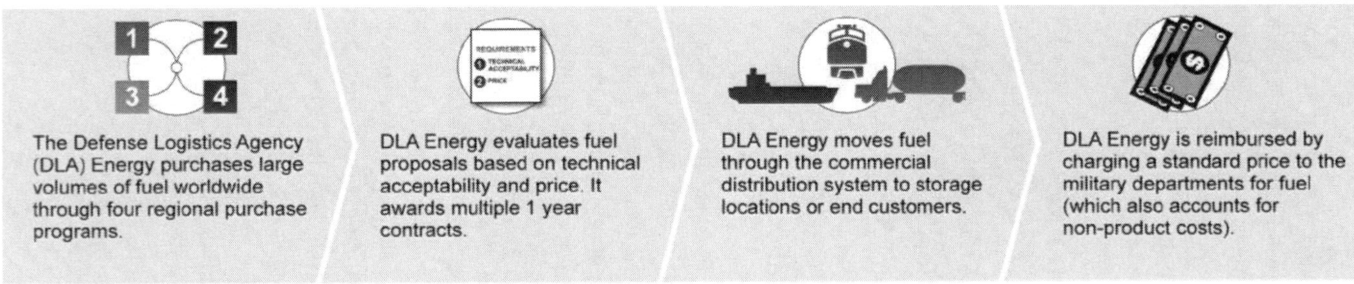

Source: GAO analysis of Department of Defense (DOD) information. | GAO-15-674.

Figure 2. Standard Process to Purchase Large Scale Volumes of Fuel.

DLA-E utilizes its Defense-wide Working Capital Fund for large-scale fuel purchases for its military and other government customers. According to DOD's Financial Management Regulation, working capital funds were established to satisfy recurring DOD requirements using a businesslike buyer-and-seller approach.[42] The fund covers DLA-E's costs for purchasing large quantities of fuel and is reimbursed through its sale of the fuel to the military at a standard price. The standard price is also based on, among other things, an estimate for non-product costs such as transportation and storage costs. The standard price is intended to remain unchanged until the next fiscal year. To simplify cost planning and budgeting, the standard price for a given fuel is the same globally.

DOD is Currently Required to Consider Whether Alternative Fuels Are Cost-Competitive with Conventional Fuels for Military Operations

As DOD seeks to purchase alternative fuels for military operations, it is required to consider whether alternative fuels are cost-competitive with conventional fuels.[43] Under the previously mentioned DOD technical fuel specifications, fuels produced through the Fischer-Tropsch and Hydroprocessed Esters and Fatty Acids production processes are approved to be used in Navy aviation and ship platforms, Air Force aviation assets, and Army tactical and combat ground vehicles and ground support equipment but not yet for Army aviation assets for a blend of up to 50 percent with conventional fuel. [44] From a technical requirements perspective, DOD can purchase and use alternative fuels produced via these approved processes for military operations. However, in conforming to the law and to departmental guidance, DOD must currently consider whether alternative fuels are cost-competitive with conventional fuels. For example, DOD may not obligate or expend funds made available for fiscal year 2015 to make a large-scale purchase of alternative fuel for operational purposes unless the fully burdened cost— that is, the commodity price of the fuel plus the total cost of all personnel and assets required to move and, when necessary, protect the fuel from the point at which the fuel is received from the commercial supplier to the point of use—of that fuel is cost-competitive with the fully burdened cost of conventional fuel.[45] However, with the requisite notice to the congressional defense committees, the Secretary of Defense may waive this limitation and the Secretary is required to notify the congressional defense committees no later than 30 days before the purchase date if DOD intends to purchase an alternative fuel for operational use that has a fully burdened cost that is in excess of 10 percent more than the fully burdened cost of conventional fuel for the same purpose.[46] A similar provision was in effect for fiscal year 2014 funds, but it did not reference the fully burdened cost of fuel, nor was the 10 percent notice requirement included.[47] DOD guidance also discusses consideration of cost with regard to alternative fuel purchases for operational purposes, and DOD's Operational Energy Strategy indicates that the department will acquire such fuels for military operations at prices that are competitive with the market price for conventional fuels. DOD has also recently issued updated guidance establishing it is DOD's policy that alternative fuels for operational purposes are purchased when cost-competitive with traditional fuels and when qualified as compatible with existing equipment and infrastructure.[48]

In December 2013, the Secretaries of the Departments of Agriculture and the Navy announced an initiative, called Farm to Fleet, which is intended to help the Department of the Navy meet its alternative fuels usage goals. Related to this initiative, DOD intends to purchase, through its regular domestic fuel purchases, Jet Propellant 5 jet and naval distillate fuels meeting DOD's technical fuel specifications and that are blended with at least 10 percent but no more than 50 percent alternative fuels— specifically biofuels—for the Department of the Navy's use in military operations. The Department of Agriculture plans, under the authority of the Commodity Credit Corporation Charter Act, to contribute up to $161 million to alternative fuel purchases to help defray some of the extra costs—which may include the costs of feedstocks—that would have caused the final alternative fuel to be more expensive than the price of conventional fuels for DOD. [49] To be eligible for the Department of Agriculture's Commodity Credit Corporation funding, the specific amounts per gallon of which are provided in DOD's fuel solicitation documents, fuel vendors have to provide an alternative fuel that was produced from an approved domestic feedstock—such as crop and tree residues, algae/algal oil, or animal waste and by-products of animal waste.

In the event of a contract award with fuel vendors providing alternative fuels, the vendors would receive separate payments from the Department of Agriculture's Commodity Credit Corporation and DOD. The Department of Agriculture's Commodity Credit Corporation would pay the incentive amount per gallon indicated in the solicitation to the fuel vendors in order to help them defray some of their costs, including domestic feedstock costs. DOD would pay the remainder of these alternative fuel vendors' prices. However, the incentive is not an additional sum paid to a fuel vendor over and above the price submitted in its proposal, but rather provides Commodity Credit Corporation funds to cover the portion of the total submitted price that exceeds the price DOD would otherwise pay. In no event would fuel vendors providing alternative fuels be paid more than the price they submitted in their proposals. We note that such an arrangement means that the cost of the alternative fuel to the federal government as a whole may be higher than the cost of conventional fuel. This is because, while DOD would be paying a price that is competitive with the price of conventional fuel, the Department of Agriculture would be paying an additional subsidy.

DOD's first attempt to purchase alternative fuels for military operations, through its large-scale fuel program, occurred in June 2014, when DOD issued a solicitation, through its regular domestic large-scale fuel purchase program for the Inland/East/Gulf Coast/Offshore region of the United States, for the purchase of Jet Propellant-5 jet and naval distillate fuels for which blended fuels with between 10 to 50 percent alternative fuels were to be considered. The solicitation listed the estimated maximum quantity of Jet Propellant-5 and naval distillate fuels as approximately 392.5 million gallons. As such, the maximum amount of biofuel to be blended into the desired fuel amount would be equivalent to approximately 39 million gallons (at 10 percent blend) to 196 million gallons (at 50 percent blend). The Department of Agriculture made available approximately $27 million in Commodity Credit Corporation funds to support successful biofuel contract awards. According to DOD officials, proposals with biofuel bids for only naval distillate fuel but not Jet Propellant-5 jet fuel were received. However, according to DOD officials, none of the submitted proposals successfully met all of the technical evaluation factors. The fuel contract awards under this solicitation were announced on February 20, 2015. According to DOD officials, none were for alternative fuels.

DOD's second attempt to purchase alternative fuels for military operations, through its large-scale fuel program, began in April 2015. At that time, DOD issued a solicitation, through its regular domestic large-scale fuel purchase program for the Rocky Mountain/West Coast/Offshore region of the United States, for the purchase of Jet Propellant-5 jet and naval distillate fuels for which blended fuels with between 10 to 50 percent alternative fuels were to be considered. The solicitation listed the estimated maximum quantity of Jet Propellant-5 and naval distillate fuels as approximately 290.6 million gallons. As such, the maximum amount of biofuel to be blended into the desired fuel amount would be equivalent to approximately 29 million gallons (at 10 percent blend) to 145 million gallons (at 50 percent blend). The Department of

Agriculture has made available approximately $66 million in Commodity Credit Corporation funds to support successful biofuel contract awards. Fuel vendors had until May 18, 2015, to submit proposals, and DOD plans to make contract awards before October 1, 2015.

According to a Navy official, the Department of Agriculture's Commodity Credit Corporation funds will not be available for DOD's regular fuel purchase programs for the Atlantic/European/Mediterranean and Western Pacific regions because those are international rather than domestic fuel purchases.

DOD HAS USED DEFENSE PRODUCTION ACT AUTHORITIES IN COLLABORATION WITH PRIVATE INDUSTRY TO PROMOTE DOMESTIC BIOFUEL PRODUCTION

Defense Production Act Allows DOD to Provide Financial Incentives to Private Firms to Meet Critical National Defense Needs

Title III of the Defense Production Act (DPA)—Expansion of Productive Capacity and Supply—generally allows military and civilian agencies to provide a variety of financial incentives to domestic firms to invest in production capabilities, so as to ensure that the domestic industrial and technological base is capable of meeting the national defense needs of the United States.[50] Use of certain Title III authorities requires a determination that, among other things, the industrial resource, material, or critical technology item is essential to national defense and U.S. industry cannot reasonably be expected to provide the capability needed in a timely manner.[51] Title III financial incentives can reduce the risks for domestic suppliers associated with the capitalization and investments required to establish, expand, or preserve production capabilities. According to DOD officials, the focus of Title III is to establish commercially viable industrial capabilities that will continue to prosper after federal government assistance ends. Funding for Title III projects comes from appropriations for DPA purchases, DOD components, or other federal agencies.

DOD Has Used DPA Title III Authority for Two Biofuel Production Projects

Bio-Synthetic Paraffinic Kerosene Project

DOD first used Title III authority in relation to alternative fuels in 2010, for the purpose of producing Bio-Synthetic Paraffinic Kerosene, an alternative jet and naval distillate fuel made from the Hydroprocessed Esters and Fatty Acids production process. Alternative fuels made from this process can meet DOD's technical fuel specifications when blended with conventional fuels. The modified biorefinery resulting from this project is to produce Bio-Synthetic Paraffinic Kerosene fuels and other co-products from natural oils, fat, and grease feedstocks via the Hydroprocessed Esters and Fatty Acids production process.

According to DOD officials, the project originated from the Department of Defense Appropriations Act for Fiscal Year 2010. The explanatory statement for that act listed Bio-Synthetic Paraffinic Kerosene Production among the DPA projects for that year.[52] In September 2009, the Air Force, as Executive Agent for the DPA Title III program, issued a Request for Information inviting the private sector to provide information about establishing manufacturing capability for Bio-Synthetic Paraffinic Kerosene fuel derived from renewable biomass feedstock sources.[53] According to DOD officials, only one private company responded to the request. In December 2010, the Under Secretary of Defense for Acquisition, Technology, and Logistics issued a determination that (1) the industrial resource or technology item of Bio-Synthetic Paraffinic Kerosene fuel is essential for national defense, and (2) U.S. industry cannot reasonably be expected to provide this fuel in a timely manner without action under the Defense Production Act. This written determination was submitted to relevant congressional committees.[54] Before awarding the Bio-Synthetic Paraffinic Kerosene production agreement, DOD made a second attempt to identify other private companies with expertise in this area. With no additional responses, DOD entered into a technology investment agreement in September 2012 with the sole private company that had previously responded to the Request for Information.

According to DOD officials, the modified biorefinery project began in September 2012 and the biorefinery is expected to be completed and operational in September 2015. They stated that it has an end goal of producing alternative fuels and co-products in a volume between 20 and 28 million gallons per year as of when DPA Title III assistance ends, and will have the capability to blend the alternative fuel with conventional petroleum fuel for its customers. DOD's financial contribution for this project comprises approximately $4 million of its total cost, with the awarded private company paying the remainder. A commercial airline has announced that it has entered into an agreement with the biofuel refinery to purchase 15 million gallons of the jet fuel produced by this project over 3 years. Also, the biofuel refinery announced that it has a strategic partnership with a fuel distributor that supplies aviation fuel. DOD officials noted that, although the department has not entered into any agreement to purchase alternative fuel from this modified biorefinery, the biofuel refinery would be able to compete for a fuel contract with DOD via DLA-E's existing large-scale fuel purchase process.

Advanced Drop-in Biofuels Production Project

DOD's second use of Title III authority in relation to alternative fuels began in 2012 with the Advanced Drop-in Biofuels Production Project. The goal of the project is to establish one or more domestic integrated biofuels production enterprise capable of annually producing at least 10 million gallons of alternative jet and/or naval distillate fuel that can meet DOD's technical fuel specifications. This enterprise would include feedstock acquisition and logistics, conversion facilities (Integrated Biorefineries), and fuel blending, transportation, and logistics. The effort would include the design, construction or retrofit, validation, qualification, and operation of a domestic commercial-scale integrated biofuels production enterprise.

In June 2011, the Department of Agriculture, Department of Energy and Department of the Navy signed a memorandum of understanding that initiated a cooperative effort to assist in the development and support of a sustainable commercial biofuels industry. This occurred in response to the president's March 2011 Blueprint for a Secure Energy Future, which challenged the Secretaries of these three departments to investigate how they might work together to speed the development of drop-in biofuels substitutes for diesel and jet fuel. The Blueprint noted that competitively priced drop-in biofuels could help meet the fuel needs of the Navy, as well as the commercial aviation and shipping sectors. The memorandum of understanding explained that given the current economic environment, significant start-up risks, and the competitive barriers posed by the firmly established conventional fuels market, private industry would not assume all of the uncertainty and risk associated with providing a commercially viable production capability for drop-in biofuels. Accordingly, it was necessary for the federal government to cooperate with industry to create a strong demand signal and to make targeted investments to achieve the necessary alternative fuels production capacity. The stated objective was to construct or retrofit multiple domestic commercial- or pre-commercialscale advanced drop-in biofuel plants and refineries. Specific characteristics for these facilities include the capability to produce biofuels meeting DOD's technical fuel specifications at a price that would be competitive with conventional fuel, and that they would cause no significant impact on the supply of agricultural commodities for the production of food. Under the memorandum of understanding, the three departments stated their intentions to equally contribute funding over a period of 3 years. The Departments of Energy and the Navy plan to apply their funds through the DPA. The Department of Agriculture plans to provide its contribution via the Commodity Credit Corporation funds, as discussed above.

In August 2011, the Air Force, as Executive Agent for the DPA Title III program, issued a Request for Information to the private sector to obtain information related to advanced drop-in hydrocarbon biofuels production, including the technical, manufacturing, and market barriers to establishing a viable business for producing biofuels. According to DOD officials, an interagency team was formed to review the responses and use the findings as guidance to develop the requirements for the biofuel project. In June 2012, DOD announced the initiation of and a solicitation for the Advanced Drop-in Biofuels Production Project, which would provide awards for biofuels production facilities over two phases.[55] Phase I awards would be for planning and preliminary designs for biofuel production facilities; and Phase II awards would be for constructing, commissioning, and performance testing of biofuel production facilities. In January 2013, the Under Secretary of Defense for Acquisition, Technology and Logistics issued a determination that (1) an advanced drop-in biofuels production capability is essential to the national defense; and (2) without action under DPA authority, U.S. industry could not reasonably be expected to provide the capability in a timely manner. This written determination was submitted to relevant congressional committees. In May and June 2013, DOD selected four private companies to receive Phase I awards totaling $20.5 million, with private industry contributing funds for the remainder of the Phase I costs (at least 50 percent).[56]

Only Phase I awardees were eligible to apply for the Phase II awards. In August 2014, three of the four Phase I awardees received Phase II awards totaling $210 million, with private industry contributing funds for the remainder of the costs (which are to be more than 50 percent). Phase II awardees are currently performing activities in preparation for constructing their biorefinery facilities, including conducting environmental analyses and securing financing. According to DOD officials, DOD will monitor the Phase II awardees by conducting biweekly teleconferences, quarterly status reporting updates, and site visits. In general, monitoring activities will monitor factors such as whether there are any changes to the company's project scope, implementation, or timelines; and if the company's amount of spending correlates to how much of the biofuel production facility has been completed. As shown in Table 1 below, the Phase II awardees will be making alternative fuel from different production processes and deriving it from various feedstock sources. According to DOD program officials, the Advanced Drop-in Biofuels Production Project should provide, between 2017 to 2018, production capacity for about 106 million gallons per year of alternative jet and naval distillate fuels that meet DOD's technical fuel specifications and are available at a price that is competitive with that of conventional fuels.

DOD applied $100 million in fiscal year 2012 procurement funds to this project. For fiscal year 2013, the explanatory statement for the Consolidated and Further Continuing Appropriations Act, 2013, listed $60 million for this project.[57] The National Defense Authorization Act for Fiscal Year 2013 provided that amounts made available to DOD under the DPA for fiscal year 2013 for biofuels production could not be obligated or expended for the construction of a biofuel refinery until matching contributions were received from the Department of Energy and equivalent contributions from the Department of Agriculture.[58] For fiscal years 2014 and 2015, the Department of Energy received authorization to transfer up to $45 million each year for DPA purposes.[59] The Department of Energy has contributed those funds to the DPA fund for this project. For the Department of Agriculture's equivalent contribution, it has committed to expenditures of Commodity Credit Corporation funding through the initiative described above. Two commercial airlines have announced they are entering into fuel purchase agreements with two of the Phase II awardees. The third Phase II awardee, according to DOD officials, is also in talks with potential non-DOD customers. DOD officials noted that, although the department has not entered into any agreement to purchase alternative fuels from these private companies, they would be able to compete for a contract with DOD via DLA-E's existing large-scale fuel purchase process.

Table 1. Summary of Activities for Advanced Drop-In Biofuels Production Project Phase II Companies (Awarded in August 2014)

	Biofuels company #1	**Biofuels company #2**	**Biofuels company #3**
Projected volumes of all alternative fuel to be produced per year (in gallons)	83 million gallons per year	10.6 million gallons per year	12.3 million gallons per year
Type(s) of alternative fuel to be produced	•Diesel •Naval distillate	•Jet (for both commercial aviation and military use)	•Jet (for both commercial aviation and military use) •Diesel
Alternative fuel production process technology	Hydroprocessed Esters & Fatty Acids	Fischer-Tropsch	Fischer-Tropsch
Feedstock source	Fats, oils, and grease	Municipal solid waste	Woody biomass from forest residue
Agreements with non-DOD customers	Under development	International airline	Domestic airline
Expected alternative fuel availability date	2018	2018	2017

Source: GAO analysis of Department of Defense (DOD) information. | GAO-15-674

Agency Comments

We are not making any recommendations in this report. We provided DOD with a draft of this report for review. DOD provided technical comments on our findings, which we have incorporated where appropriate.

Sincerely yours,

Zina D. Merritt
Director
Defense Capabilities and Management

Appendix I: Military Department Guidance on Alternative Fuels

Details about each military department's guidance as related to alternative fuels include the following.

- The Department of the Navy's 2010 Energy Program for Security and Independence[1] guidance states that increasing its use of alternative energy—including alternative fuels—-will help Naval forces have the ability, among other things, to protect and deliver sufficient energy to meet operational needs, and will enable Naval forces to rely on energy resources that are not subject to supply disruptions.[2] The guidance sets a goal of deriving 50 percent of total Department of the Navy energy consumption from alternative sources by 2020, which, according to Navy estimates, would require using about 336 million gallons of alternative fuels (both naval distillate and jet fuels) annually by 2020. The Department of the Navy purchased more than 1 billion gallons of petroleum-based jet and naval distillate fuel in fiscal year 2014.[3] This was about 90 percent of all of the Department of the Navy's fiscal year 2014 petroleum and other fuel product purchases.[4] In addition to setting quantitative goals, the guidance established a goal of demonstrating (which the Department of the Navy completed in July 2012) and deploying the Great Green Fleet (by 2016)—that is, ships and aircraft fueled by alternative fuels and other alternative energy sources or utilizing other energy conservation measures.[5]
- The Department of the Air Force's 2013 U.S. Air Force Energy Strategic Plan includes a goal of increasing, to 50 percent of total consumption, the use of cost-competitive drop-in alternative jet fuel blends for non-contingency operations by 2025.[6] The Department of the Air Force purchased more than 2 billion gallons of petroleum-based jet fuel in fiscal year 2014.[7] This was about 97 percent of all of the Department of the Air Force's fiscal year 2014 petroleum and other fuel product purchases.[8] The plan indicates that using alternative jet fuels could help to diversify the types of energy and obtain the quantities of energy that are needed to perform the Air Force's missions, which are currently "heavily dependent" upon petroleum and petroleum-derived fuels, thereby posing significant strategic and security vulnerabilities.

- The Department of the Army uses jet fuel in its tactical and ground combat vehicles, aircraft, and other ground support equipment (such as generators) and engages in efforts to test and approve alternative jet fuel made from different production processes for use in these platforms. However, the Army does not have specific alternative fuel usage goals in its energy guidance. The Department of the Army's 2015 Energy Security and Sustainability Strategy includes the strategic goals, among others, of: optimizing use and assuring access. [9] To accomplish these goals, the Department of the Army plans to minimize overall energy demand and improve efficiency, while securing access to renewable/alternative energy sources to diversify and expand its resource supply, among other actions. The Department of the Army purchased more than 350 million gallons of petroleum-based jet fuel in fiscal year 2014.[10] This was about 76 percent of all of the Department of the Army's fiscal year 2014 petroleum and other fuel product purchases.[11]

APPENDIX II: DETAILED QUANTITY AND COST DATA OF MILITARY DEPARTMENTS' ALTERNATIVE AND CONVENTIONAL FUEL PURCHASES

Tables 2 and 3 show detailed quantity and cost data by fiscal year of the alternative and conventional petroleum jet and naval distillate fuels that the military departments purchased from fiscal years 2007 through 2014.

Table 2. Quantities and Cost of Alternative Jet and Naval Distillate Fuels Purchased by the Military Departments, Fiscal Years 2007 through 2014

Fiscal year	Quantity(gallons in thousands)	Cost (dollars in thousands and inflation-adjusted)
2007	315.0	$1,219.1
2008	395.0	1,703.0
2009	261.5	17,704.6
2010	385.0	14,433.4
2011	15.5	851.7
2012	514.4	15,954.0
2013	76.1	3,092.2

Table 2. (Continued)

Fiscal year	Quantity(gallons in thousands)	Cost (dollars in thousands and inflation-adjusted)
2014	71.5	3,610.7
Total	2,034	58,568.8

Source: GAO analysis of DOD data. | GAO-15-674

Note: Totals may not sum due to rounding. The quantity and cost amounts include: Jet Propellant-8 and Jet Propellant-5 products and Naval Distillate fuel (known as F-76). The cost amounts have been adjusted for inflation to fiscal year 2015 dollars using the gross domestic product price index.

Table 3. Quantities and Cost of Conventional Jet and Naval Distillate Fuels Purchased by the Military Departments, Fiscal Years 2007 through 2014

Fiscal year	Quantity (gallons in billions)	Cost (dollars in billons and inflation-adjusted)
2007	4.4	$11.1
2008	4.3	15.1
2009	4.2	10.1
2010	4.2	13.0
2011	4.3	16.0
2012	3.9	15.3
2013	3.4	13.7
2014	3.3	12.9
Total	32.0	107.2

Source: GAO analysis of DOD data. | GAO-15-674

Note: Totals may not sum due to rounding. The quantity and cost amounts include: Jet Propellant-8, Jet Propellant-5, Jet A, Jet A-1 products and Naval Distillate fuel (known as F-76). The cost amounts have been adjusted for inflation to fiscal year 2015 dollars using the gross domestic product price index.

APPENDIX III: FUEL PROPERTIES

Before any alternative fuel can be used in military operations, it is tested and approved to meet unique safety and performance standards. Listed below are examples of fuel properties important to these standards. Requirements for alternative fuels are set out in the relevant DOD technical fuel specification documents.[1]

- Flash point – A liquid fuel's flash point indicates the temperature at which existing vapors will combust, or ignite. Fuels with higher flash points contribute to a less flammable, less hazardous fuel for better safety and combat survivability. Jet fuel (specifically Jet Propellant-5) used on military ships is required to have a substantially higher flash point than other jet fuel for safety reasons since this fuel is stored in large quantities on aircraft carriers and other vessels.
- Energy density[2] – The amount of potential energy in a given measurement of liquid fuel is its energy density. This property is important because it affects the distance for which military equipment and platforms using a particular liquid fuel can operate before needing to refuel; the amount of fuel that must be carried to achieve a given refueling range; and the logistics support—such as more or bigger storage tanks for a less energy-dense fuel—needed for a particular fuel.
- Freezing point – The point at which a liquid fuel freezes affects how it behaves at low temperatures. For example, a liquid fuel's freezing point can potentially have an effect on certain long-range, high-altitude missions during which extreme cold temperatures are encountered. Even though a fuel may not freeze completely at the "freezing point" and below, solid material floating in the fuel can block filters and interfere with fuel system operation. At low temperatures, the fuel may also have too high a viscosity (see definition below) for fuel systems to operate properly. In addition, this can affect the cold starting ability of engines.
- Thermal stability – A liquid fuel's thermal stability refers to its ability to accommodate increased temperatures without compromising its chemical integrity. For example, thermal stability can affect how cool the components coming in contact with the liquid fuel can remain. Also, it can affect the rate of deposits forming when the fuel temperature is elevated; these deposits on components can affect their performance, such as reducing the fuel flow through fuel filters.
- Lubricity – A liquid fuel's lubricity refers to its effectiveness in reducing friction between moving parts in equipment such as pumps and fuel control units.

- Viscosity – A liquid fuel's viscosity—which is critical to proper equipment operations—is a measure of its internal resistance to motion or flow.

APPENDIX IV: ALTERNATIVE FUEL TESTING PROTOCOLS

The Departments of the Navy[1], Air Force, and Army test alternative fuels to ensure that they can be used in and on tactical and combat ground vehicles and ground support equipment, ships, aircraft, and fuel distribution systems. In general, the military departments use the test protocols listed below, which are similar steps to those used in evaluating whether to include prospective alternative fuels in the commercial jet fuel standard issued by ASTM International.[2]

- Specification properties – This laboratory testing protocol evaluates how a prospective alternative fuel's basic chemical and physical properties—such as its freezing and flash points—compare with the baseline properties of conventional jet or naval distillate petroleum fuel. These required properties are outlined in the associated technical fuel specification documents.
- Fit for purpose – This laboratory testing protocol involves evaluating additional properties that are inherent to conventional jet or naval distillate petroleum fuel, such as how compatible a prospective alternative fuel is with specific metallic and non-metallic materials and various additives that, among other things, inhibit corrosion and dissipate static.
- Component/Rig testing - This testing protocol involves evaluating how a prospective alternative fuel performs in major components found in the military department's ground vehicles and support equipment, ships, and aircraft. Examples of these components include injectors and the section of an engine where combustion occurs.
- Full scale testing – This testing protocol involves evaluating how a prospective alternative fuel performs in engines of ground vehicles and support equipment, ship and aircraft engines, auxiliary power units, and fuel handling systems.
- Platform testing - This testing protocol involves evaluating how a prospective alternative fuel performs when different types of ground

vehicles and support equipment, ships, aircraft, and fuel support equipment run on the fuel. Typically, these equipment and platforms are the military department's assets that are running on alternative fuel in settings and under conditions that mimic environments where military operations may occur.

End Notes

[1] Operational energy is defined in section 2924 of Title 10, U.S. Code.

[2] The Department of the Navy includes two military services—the Navy and the Marine Corps.

[3] Department of Defense, *Energy for the Warfighter: Operational Energy Strategy* (May 2011).

[4] For purposes of this report, alternative fuels are focused on use only for operational energy needs and thus defined as liquid fuels whose use does not necessitate any modifications to equipment (such as platform engines and fuel distribution infrastructure)—and which are thus termed as "drop-in" liquid fuels—that are (1) derived from non-petroleum feedstocks, including renewable biomass (such as crop and tree residues, algae, or separated municipal solid waste) and some nonrenewable sources (such as natural gas or coal); and (2) used in platforms and equipment that are intended for use in or support of military operations. DOD currently intends to use drop-in alternative fuels blended with conventional petroleum-based fuel without having to modify current equipment, platforms, or fuel distribution infrastructure. Most development activities are focused on alternative fuels derived from renewable sources and are known as biofuels.

[5] Department of Defense, Assistant Secretary of Defense for Operational Energy Plans and Programs, *Department of Defense Alternative Fuels Policy for Operational Platforms* (July 5, 2012).

[6] According to DOD, anti-access refers to the actions and capabilities designed to prevent an opposing force from entering an operational area and area denial refers to the actions and capabilities designed to limit an opposing force's freedom of action within an operational area.

[7] Department of Defense, *Quadrennial Defense Review* (Mar. 4, 2014).

[8] The Defense Production Act generally provides the authority to, among other things, expedite and expand the supply of critical resources—including energy supply—from the U.S. industrial base to support the national defense. *See generally* Pub. L. No. 81-774 (1950) (codified as amended at 50 U.S.C. app. §§ 2061-2172).

[9] According to DOD guidance, using Jet Propellant-8 as the primary fuel in military operations is beneficial for several reasons, including that a single fuel is easier to manage than multiple fuels, allowing the functions of fuel storage, transportation, and distribution to be tailored for maximum efficiency, and can lessen the possibility of dispensing the wrong fuel. *See* Joint Chiefs of Staff, Joint Pub. 4-03, *Joint Bulk Petroleum and Water Doctrine* at I-1 (Dec. 9, 2010).

[10] Requirements for Jet Propellant-8 are covered by Military Detail Specification MIL-DTL-83133H (Dec. 24, 2013). Requirements for Jet Propellant-5 are covered by Military Detail Specification MIL-DTL-5624V (July 11, 2013).

[11] Jet A is predominately used in the United States, while Jet A-1 is mostly used outside the United States.

[12] ASTM International, formerly known as the American Society for Testing and Materials, develops and delivers international voluntary consensus standards. ASTM Standard D1655-14c is the standard specification for aviation turbine fuels of Jet A and Jet A-1.

[13] The one grade of naval distillate fuel, F-76, is covered by Military Detail Specification MIL-DTL-16884N (Apr. 22, 2014).

[14] The International Organization for Standardization has a marine fuel standard, but according to Navy officials, it does not meet the Navy's fuel requirements, such as the need to store the fuel for longer periods of time.

[15] Department of Defense Directive 4180.01, *DOD Energy Policy*, para. 3.b (Apr. 16, 2014).

[16] This position was recently merged with the former Deputy Under Secretary of Defense (Installations & Environment) position to create the Assistant Secretary of Defense for Energy, Installations, and Environment. *See* Carl Levin and Howard P. "Buck" McKeon National Defense Authorization Act for Fiscal Year 2015, Pub. L. No. 113-291, § 901(f) (2014) (amending 10 U.S.C. § 138(b)(9)). Many of the responsibilities of the Assistant Secretary of Defense for Operational Energy Plans and Programs were transferred from section 138c of Title 10, U.S. Code, to section 2926. *See id.* § 901(g).

[17] This responsibility is to be carried out in accordance with Department of Defense Directive 4140.25, *DOD Management Policy for Energy Commodities and Related Services* (Apr. 12, 2004). This directive was reissued as Department of Defense Instruction 4140.25 on June 25, 2015.

[18] Department of Defense Instruction 4140.25, *DOD Management Policy for Energy Commodities and Related Services*, paras. 3.d, 3e, (June 25, 2015). This instruction reissues Department of Defense Directive 4140.25 as a Department of Defense Instruction.

[19] *Department of Defense Alternative Fuels Policy for Operational Platforms* (July 5, 2012).

[20] Department of Defense, *Energy for the Warfighter: Operational Energy Strategy* (May 2011), and *Operational Energy Strategy: Implementation Plan* (March 2012).

[21] These energy guidance documents include: Department of the Navy, *Department of the Navy's Energy Program for Security and Independence* (October 2010); Department of the Air Force, *U.S. Air Force Energy Strategic Plan* (March 2013); and Department of the Army, *Energy Security & Sustainability (ES^2) Strategy* (May 1, 2015).

[22] This demonstration, known as the Great Green Fleet, occurred with a group of ships and aircraft fueled by alternative fuels in an operational environment that was part of a larger, biennial multinational maritime exercise, known as the Rim of the Pacific exercise.

[23] GAO, *Alternative Jet Fuels: Federal Activities Support Development and Usage, but Long-term Commercial Viability Hinges on Market Factors,* GAO-14-407 (Washington, D.C.: May 7, 2014).

[24] *See generally* Pub. L. No. 81-774 (1950) (codified as amended at 50 U.S.C. app. §§ 2061-2172). As originally enacted, the Defense Production Act of 1950 principally addressed the availability of goods and services necessary to meet defense needs. Amendments to the Act allow the funds contributed under its authority to be used for energy supply, emergency preparedness, and critical infrastructure protection.

[25] The term "national defense" as used in the DPA means programs for military and energy production or construction, military or critical infrastructure assistance to any foreign nation, homeland security, stockpiling, space, and any directly related activity. The term includes emergency preparedness activities conducted pursuant to certain statutes and critical infrastructure protection and restoration. 50 U.S.C. app. § 2152(14).

[26] *See generally* 50 U.S.C. app. §§ 2091-2094.

[27] *See* § 2093(a)(5). The determination requirement in section 2093(a)(5) was amended in September 2014 to prohibit delegation and add to the information required to be in the determination. *See* DPA Reauthorization Act of 2014, Pub. L. No. 113-172, § 4(a)(1) (2014).

[28] *See* Department of Defense Directive 4400.01E, *Defense Production Act Programs*, paras. 4.1.2, 4.1.10 (Oct. 12, 2001) (certified as current as of Sept. 14, 2007). According to DOD, the Under Secretary of Defense for Acquisition, Technology, and Logistics is assisted by the Deputy Assistant Secretary of Defense (Manufacturing & Industrial Base Policy) and the Program Director of the Defense Production Act Title III Program.

[29] These conditions include the following: that the fuels are "drop-in," in the sense that they can replace conventional fuels without the need to modify existing military equipment and infrastructure; that the fuels are able to support an expeditionary, globally deployed force; that there is consideration of consequences, such as higher food prices; and that lifecycle greenhouse gas emissions are less than or equal to those from conventional fuels such that, according to DOD officials, production of the alternative fuels is consistent with section 526 of the Energy Independence and Security Act of 2007, Pub. L. No. 110- 140 (2007).

[30] This cost information does not include any non-fuel costs—such as employee salaries, laboratory equipment time, or contracts with third parties to conduct specific tests—the military departments incurred related to testing alternative fuels or other research and development activities.

[31] The Rim of the Pacific exercise has taken place since 1971 in and around the Hawaiian Islands.

[32] This cost information reflects both the fuel and non-product costs—such as storage and transportation costs—that the military departments paid.

[33] Requirements for synthesized jet fuel for military-grade Jet Propellant-8 is covered in Military Detail Specification MIL-DTL-83133H (Dec. 24, 2013). Requirements for synthesized jet fuel for military grade Jet Propellant-5 is covered in Military Detail Specification MIL-DTL-5624V (July 11, 2013). The one grade of synthesized naval distillate fuel, F-76, is covered by Military Detail Specification MIL-DTL-16884N (Apr. 22, 2014).

[34] ASTM International, formerly known as the American Society for Testing and Materials, develops and delivers international voluntary consensus standards. ASTM Standard D7566 covers the manufacture of jet fuel containing blends of conventional and synthesized hydrocarbons (those not derived from petroleum hydrocarbons) for commercial use.

[35] Under the Fischer-Tropsch process, biomass (such as switchgrass or wood waste), coal, or natural gas is processed through gasification into synthesis gas, and that gas is then converted to synthetic liquid fuels. Under the Hydroprocessed Esters and Fatty Acids process, renewable oil (for example, vegetable oils, animal fat, waste grease, or algal oil) is processed using hydrogen treatment (hydroprocessing) to yield a hydrocarbon fuel in the distillation range of jet fuel and diesel.

[36] Under the Alcohol to Jet process, biomass or municipal solid waste is processed through fermentation to create alcohols that are further processed through dehydration into small hydrocarbon molecules and then reassembled into jet and diesel fuels.

[37] Under the Synthesized Iso-Paraffins production process, sugars (from sources such as the sugarcane plant) are directly fermented into hydrocarbons that are then processed using hydrogen treatment to yield jet and diesel fuel blending components. Under the Catalytic Hydrothermolysis production process, renewable oils are processed in multiple ways including heat and hydrotreating to yield hydrocarbon jet and diesel fuels.

[38] Under the Hydroprocessed Depolymerized Cellulosic production process, renewable biomass is broken down into smaller molecules, then processed in multiple ways including hydrogen treatment to yield jet, diesel, and other fuels.

[39] Air Force officials pointed out that expertise to run testing and approval efforts is still in place.

[40] The Defense Logistics Agency Energy is a primary-level field activity of the Defense Logistics Agency that is responsible for providing comprehensive energy solutions to DOD and other government agencies.

[41] According to DOD officials, the military departments can also buy fuels outside this process, but most of the fuel for military operations is purchased by DLA-E.

[42] Department of Defense Financial Management Regulation 7000.14-R, Volume 2B, Chapter 9, para. 090102 (Dec. 2014).

[43] *See* Carl Levin and Howard P. "Buck" McKeon National Defense Authorization Act for Fiscal Year 2015, Pub. L. No. 113-291, § 316 (2014); National Defense Authorization Act for Fiscal Year 2014, Pub. L. No. 113-66, § 315 (2013).

[44] According to Department of the Army officials, the testing process for alternative fuels made from these two production processes in Army aviation assets is complete; however, they stated that the review of the test results in order to decide whether to approve the use of these fuels in Army aviation assets is ongoing.

[45] Pub. L. No. 113-291, § 316. Section 316 defines operational purposes as for purposes of conducting military operations, including training, exercises, large scale demonstrations, and moving and sustaining military forces and military platforms. It does not include research, development, testing, evaluation, fuel certification, or other demonstrations. § 316(d)(2).

[46] The conferees expressed their expectation in the Joint Explanatory Statement that DOD not use a unique federal subsidy to buy or purchase down the cost of fuel so it falls below the 10 percent threshold over the fully-burdened market price of traditional fuels available for the same purpose. *See* 160 Cong. Rec. H8677 (daily ed. Dec. 4, 2014).

[47] Pub. L. No. 113-66, § 315. Section 315 similarly excluded research, development, testing, evaluation, fuel certification, or other demonstrations from the definition of operational purposes. *See* § 315(c)(3).

[48] Department of Defense Instruction 4140.25, *DOD Management Policy for Energy Commodities and Related Services,* para. 3.d. (June 25, 2015).

[49] The Commodity Credit Corporation Charter Act, Pub. L. No. 80-806, § 2 (1948), codified as amended at 15 U.S.C. 714 *et seq.*, incorporated the Commodity Credit Corporation within the Department of Agriculture for the purpose of stabilizing, supporting, and protecting farm income; assisting in the maintenance of balanced and adequate supplies of agricultural commodities, products thereof, foods, feeds, and fibers; and facilitating the orderly distribution of such commodities. The Department of Agriculture, through the Commodity Credit Corporation, is authorized, among other things, to increase the domestic consumption of agricultural commodities (other than tobacco) by expanding or aiding in the expansion of domestic markets or by developing or aiding in the development of new and additional markets, marketing facilities, and uses for such commodities. Under the Memorandum of Understanding between the Department of the Navy, Department of Energy and Department of Agriculture, it is this authority that the Department of Agriculture is exercising in its efforts to support the development of a more robust alternative fuels industry.

[50] *See generally* 50 U.S.C. app. §§ 2091-2094. The term "national defense" as used in the DPA means programs for military and energy production or construction, military or critical infrastructure assistance to any foreign nation, homeland security, stockpiling, space, and

any directly related activity. The term includes emergency preparedness activities conducted pursuant to certain statutes and critical infrastructure protection and restoration. 50 U.S.C. app. § 2152(14).

[51] *See* § 2093(a)(5).

[52] 155 Cong. Rec. H15216-17 (daily ed. Dec. 16, 2009); *see also* Pub. L. No. 111-118, § 1014 (2009).

[53] The Bio-Synthetic Paraffinic Kerosene fuel production technology was developed with funding from the Defense Advanced Research Project Agency.

[54] The determination and related materials were provided to the House Committee on Financial Services and the Senate Committee on Banking, Housing, and Urban Affairs.

[55] The goal of the project was to establish one or more complete domestic value chains capable of producing drop-in replacement biofuels, including feedstock production and logistics, conversion facilities (integrated biorefineries), and fuel blending, transportation, and logistics.

[56] Each of the four private companies' actual contribution for Phase I was more than 50% of the total cost.

[57] *See* 159 Cong. Rec. S1456 (daily ed. Mar. 11, 2013). Section 4 of the appropriations act provided that the explanatory statement would have the same effect with respect to allocation of funds and implementation of the act as if it were a joint explanatory statement of a committee of conference. Pub. L. No. 113-6, § 4 (2013).

[58] Pub. L. No. 112-239, § 315 (2013).

[59] *See* Consolidated Appropriations Act, 2014, Pub. L. No. 113-76, 128 Stat. 5, 164-65 (2014); Consolidated and Further Continuing Appropriations Act, 2015, Pub. L. No. 113- 235, 128 Stat. 2130, 2312-13 (2014).

End Notes for Appendix I

[1] Department of the Navy, *Department of the Navy's Energy Program for Security and Independence* (October 2010).

[2] The Department of the Navy includes two military services—the Navy and the Marine Corps.

[3] This amount includes all fuels classified as jet fuels—including but not limited to Jet Propellant-8, Jet Propellant-5, Jet A, and Jet A-1—and the naval distillate fuel classified as F-76.

[4] These purchased products include fuels—such as, aviation gasoline and jet fuels—and other fuel products—such as lube oils and residuals.

[5] This demonstration, known as the Great Green Fleet, occurred with a group of ships and aircraft fueled by alternative fuels in an operational environment that was part of a larger, biennial multinational maritime exercise, known as the Rim of the Pacific exercise.

[6] Department of the Air Force, *U.S. Air Force Energy Strategic Plan* (March 2013).

[7] This amount includes all fuels classified as jet fuels—including but not limited to Jet Propellant-8, Jet Propellant-5, Jet A, and Jet A-1.

[8] These purchased products include fuels—such as, aviation gasoline and jet fuels—and other fuel products—such as lube oils and residuals.

[9] Department of the Army, *Energy Security & Sustainability (ES^2) Strategy* (May 1, 2015).

[10] This amount includes all fuels classified as jet fuels—including but not limited to Jet Propellant-8, Jet Propellant-5, Jet A, and Jet A-1.

[11] These purchased products include fuels—such as, aviation gasoline and jet fuels—and other fuel products—such as lube oils and residuals.

End Notes for Appendix III

[1] Requirements for synthesized jet fuel for military-grade Jet Propellant-8 are covered in Military Detail Specification MIL-DTL-83133H (Dec. 24, 2013). Requirements for synthesized jet fuel for military-grade Jet Propellant-5 are covered in Military Detail Specification MIL-DTL-5624V (July 11, 2013). The one grade of synthesized naval distillate fuel, F-76, is covered by Military Detail Specification MIL-DTL-16884N (Apr. 22, 2014).

[2] Energy density is evaluated for both the mass and volume of fuel required.

End Notes for Appendix IV

[1] The Department of the Navy includes two military services—the Navy and the Marine Corps.

[2] ASTM International, formerly known as the American Society for Testing and Materials, develops and delivers international voluntary consensus standards. ASTM Standard D7566 covers the manufacture of jet fuel containing blends of conventional and synthesized hydrocarbons (those not derived from petroleum hydrocarbons) for commercial use.

In: Alternative Fuel Investments …
Editor: Loren Murphy
ISBN: 978-1-63484-549-6

Chapter 2

DEPARTMENT OF DEFENSE ALTERNATIVE FUELS POLICY FOR OPERATIONAL PLATFORMS*

Sharon E. Burke

The Department of Defense's (DoD's) primary alternative fuels goal is to ensure operational military readiness, improve battlespace effectiveness, and further flexibility of military operations through the ability to use multiple, reliable fuel sources. Alternative fuel investments will be targeted to ensure forces are able to carry out operations using alternative sources of fuel that improve the reliability of our overall fuel supply. Specific considerations include: increased resilience against strategic supply disruptions; dampened effect of petroleum price volatility; increased fuel options for operational commanders; and ultimately increased expeditionary effectiveness of our joint force.

It should be stressed that operational military readiness and battlespace effectiveness are the desired end-state, not simply the use of alternative fuels. Alternative fuels can be a means to ensure combat effectiveness, logistical flexibility and to mitigate Anti-Access/Area Denial (A2AD) effects.

All DoD investments in this area will be subject to a rigorous, merit-based evaluation, and will be reviewed as part of the annual budget certification

* This is an edited, reformatted and augmented version of a memorandum issued by the Department of Defense, Assistant Secretary of Defense for Operational Energy Plans and Programs, July 5, 2012.

process of the Office of the Assistant Secretary of Defense for Operational Energy Plans and Programs (OEPP).

DoD alternative fuels activities will be carried out primarily in the three Phases listed below.

PHASE 1: CERTIFICATION/QUALIFICATION

On an annual basis, the Tri-Service Petroleum, Oils, and Lubricants (POL) Users Group is responsible for providing a harmonized certification/ qualification plan that minimizes costs through technical review, coordination, and collaboration across the Services. Included in the plan will be recommendations on whether DoD should initiate new certification/ qualification activities for an alternative fuel source/production pathway (including a fuel blend). To assist in making this recommendation, the Tri-Service POL Users Group will review the most current report from the Department of Energy (DOE) to determine whether any new production pathways, aside from those which have already been or are in the process of being certified for all operational platforms, will likely be commercially competitive with conventional fuels within the next ten years.

The plan should take the following into consideration:

- Potential for drop-in compatibility and interoperability with existing equipment and infrastructure across the Services (e.g., JP-5, JP-8, F-76, Jet A or Jet A-1 w/additives);
- Potential ability to provide significant volumes of fuel to support an expeditionary, globally deployed force, including CONUS operations, at competitive cost OR potential to be produced at an operating location; and
- Potential for compliance with Sec. 526 of the Energy Independence and Security Act (EISA) of 2007.
- Potential to leverage planned or completed work performed by industry groups involved with the certification and approval of alternative fuels.

The POL Users Group will present the plan to the Defense Operational Energy Board for recommendation to the Assistant Secretary of Defense for

Operational Energy Plans and Programs and the Director for Logistics, Joint Staff in advance of the annual Program Budget Review.

This policy does not restrict research, development, testing, and evaluation (RDT&E) of specialty fuels that enhance military capabilities, e.g. missile propulsion systems and unmanned vehicles. This policy also does not restrict RDT&E activities to identify candidate fuels (or fuel additives) for certification/qualification and field demonstration. Such activities may include, but not be limited to, market investigations, cost-benefit analyses, and laboratory tests covering physical and chemical properties, material compatibility, and performance of fuel-wetted components.

PHASE 2: FIELD DEMONSTRATION

To meet a significant technical or operational need, military departments may engage in field demonstrations that require use of a new fuel beyond the certification/qualification process. Field demonstrations would examine the use of a new fuel in an operational environment as part of a system of systems, rather than in a single specific platform.

Criteria: Such a demonstration must be of finite duration and its expected benefits must be clearly justified. The operational risks associated with not carrying out such a demonstration must also be described.

Field demonstrations may be financially supported directly from the appropriated funds of the executing military department or departments or via the Defense Working Capital Fund (DWCF). The DWCF would be used only to procure fuel for a field demonstration. Other, non-fuel expenses (e.g., resources to collect and analyze data, manpower, etc.) would be the Services' responsibility. A proposal by a Service to carry out a field demonstration solely using its own appropriated funds, without impact to the budgets of other Services, will be reviewed during OEPP's annual budget certification process. The above criteria will be evaluated in this review.

A military department proposing to carry out a field demonstration funded via the DWCF must work with the Defense Logistics Agency-Energy (DLA Energy) to produce a detailed proposal. The military department(s) with DLA Energy then must present this proposal for joint review by the Office of the Secretary of Defense (OSD) Comptroller, Logistics and Materiel Readiness, and OEPP prior to final approval to proceed with the demonstration. The above criteria will be evaluated in this review.

PHASE 3: ONGOING PURCHASES

After Phase 1 and, if necessary, Phase 2, alternative drop-in replacement fuel procured for DoDwide use and distribution within the Class III (Bulk) supply chain will compete with petroleum products under the DLA Bulk Purchase and Direct Delivery Purchase Programs. Awards will be based on the ability to meet requirements at the best value to the government, including cost.

The Department of Defense Alternative Fuels Policy for Operational Platforms is pursuant to 10 U.S.C. 138c.

INDEX

Y